W9-CSS-741

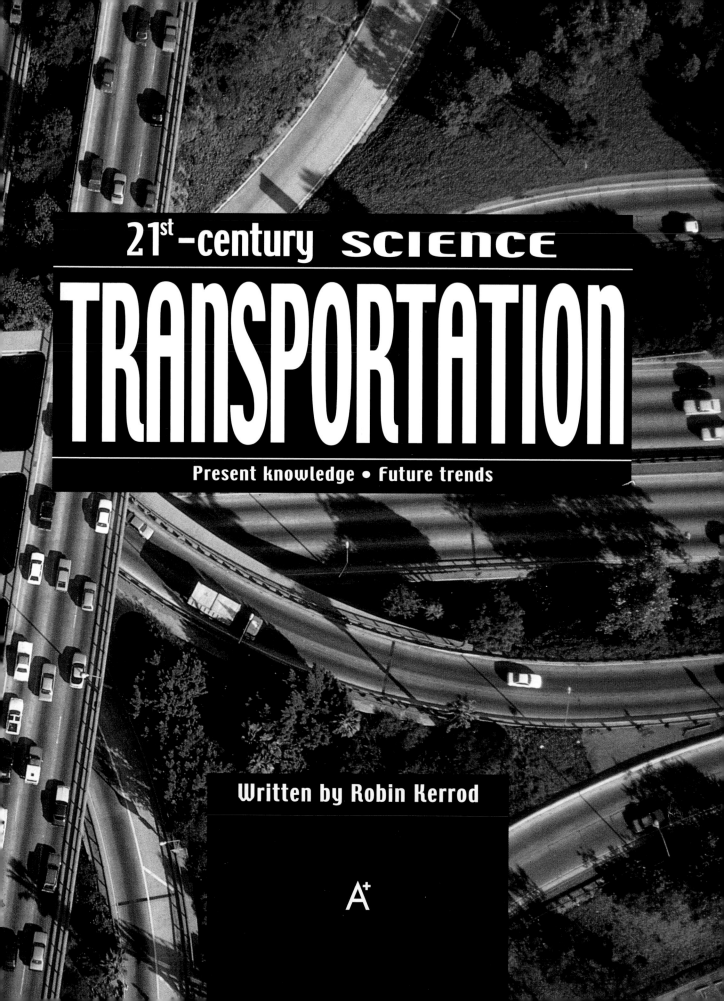

# 21st-century SCIENCE

# TRANSPORTATION

### Present knowledge • Future trends

Written by Robin Kerrod

A+

First published in 2004 by

Franklin Watts
96 Leonard Street
London EC2A 4XD

Franklin Watts Australia
45-51 Huntley Street
Alexandria
NSW 2015

**Design:** Billin Design Solutions
**Editor:** Sarah Ridley
**Art Director:** Jonathan Hair
**Editor-in-Chief:** John C. Miles
**Picture Research:** Diana Morris

**Picture credits**
Sean Aidan/Eye Ubiquitous: 9b, © Airbus: front cover br, 32t
Jean Pierre Amet/Sygma/Corbis: 22t, Ken Ayres/Ecoscene:
19 background, R. Battersby/Eye Ubiquitous: 31b , Martin
Bond/Still Pictures: 12t, James Davis Worldwide: front cover
t, front endpapers, 24t, Lionel Derimais/Sygma/Corbis: front
cover c, back cover background, 37b, Christina
Dodwell/Hutchison: 36t , Jean-Leo Dugast/Still Pictures: 11b
Peter Frischnuth/Still Pictures: 8t, Jaguar/TRH Pictures: 14t,
Joel Jensen/TRH Pictures: 20t, 21b, Ray Juno/Corbis: 17b,
Steve Kaufman/Corbis: 38t, Alex S. Maclean/Still Pictures: 4-
5 background, 13t, NASA: front cover bl, 39c,
Northrup/USAF/TRH Pictures: 34t, Sue Passmore/Eye
Ubiquitous: 28, , Thomas Raupach/Still Pictures: 26t,
Renault/TRH Pictures: 15b, Bob Rowan; Progressive
Image/Corbis: front cover bc, 30c, Alan Schein
Photography/Corbis: 27b, Paul A. Souders/Corbis: back
endpapers, 16t, 45, Jochen Tack/Still Pictures: 10c, 30t , Paul
Thompson/Eye Ubiquitous: 6, 23r , TRH Pictures: 38b,
ThyssenKrupp AG: 25c, USAF/TRH Pictures: 35b., Kevin
Wilton/Eye Ubiquitous: 29t, David Wootton/Ecoscene: 19
inset

Published in the United States by Smart Apple Media
1980 Lookout Drive, North Mankato, Minnesota 56003

Library of Congress Cataloging-in-Publication Data

Kerrod, Robin.
Transportation / by Robin Kerrod.
p. cm. — (21st century science)
ISBN 1 58340 507 0
1. Motor vehicles—Juvenile literature. I. Title. II. Series.

TL147.K463 2004
629.04—dc22          2004041601

9 8 7 6 5 4 3 2 1

# Contents

At any time of the day and night, billions of people and billions of tons of goods are on the move. They may travel in ships that are as big as floating cities, trains that are shaped like bullets, cars that navigate by satellite, or airplanes that fly higher than Mount Everest.

# Introduction

▲

*A modern, high-speed ICE (Inter-City Express) tilting train emerges from a tunnel in Germany.*

Two hundred years ago, transportation was limited. Most people did not venture very far from their homes. They traveled across land on foot or on horseback. Horse-drawn carriages were available for longer journeys, but they were hazardous because of the pot-holed roads and the risk of being stopped by bandits.

Canals were coming into use to transport heavy goods, such as coal and iron ore, as the Industrial Revolution got underway. At sea, cargo ships such as the East Indiamen were carrying cotton, silks, tea, and spices to Europe from the East. Other ships carried passengers traveling to the "New World" of America, or other countries, to establish colonies. These sailing ships were at the mercy of the winds, and they were very slow. A journey across the Atlantic could take weeks or even months.

## The revolution begins

However, in the early years of the 19th century, things began to change. In 1804, British mining engineer Richard Trevithick (1771–1833) built a self-propelled steam engine that ran along an iron track on wheels. It was the first locomotive.

But it took the genius of another British engineer, George Stephenson (1781–1848), to transform Trevithick's idea into the first successful public railroad. The Liverpool and Manchester Railway opened in 1830.

Railroad fever gripped the world, particularly in North America, where railroads accelerated the widespread settlement of the vast continent. Elsewhere, railroads opened up long-distance travel and the rapid transport of goods during the Industrial Revolution.

## The horseless carriage

Self-propelled road transportation remained a problem. Then in 1885, German engineers Gottlieb Daimler (1834–1900) and Karl Benz (1844–1929) hit on the same idea—to use a gas engine to power a carriage. The automobile was born.

By 1908, Henry Ford (1863–1947) was applying mass-production techniques to produce the legendary Model T, or "Tin Lizzie," very cheaply. He brought the automobile to the masses for the first time. People's love affair with the car began.

## The airplane

In 1903, a new form of transportation had come onto the scene. Two American brothers, Wilbur (1867–1912) and Orville (1871–1948) Wright, both bicycle mechanics, made the first powered flight. They fitted a gas engine to a glider they had built and invented the airplane.

Aviation—flying in planes—quickly became popular. The potential of the airplane as a weapon was also soon realized. During World War I (1914–18) and afterward, improvements in aircraft design came fast and furious. In the 1930s, the forerunners of the modern airliner came into operation—the Boeing 247 (1933) and the Douglas DC–3, known as the Dakota (1935).

## Onward and upward

As fighters and bombers, aircraft played a decisive role in World War II (1939–45). By the war's end, an entirely new kind of plane was streaking through the air. It did not use propellers to power it forward like all previous aircraft. It was jet-propelled, which meant it used a stream of hot gases instead.

In 1947, U.S. pilot Chuck Yeager (1923–) passed another aviation milestone and flew a rocket plane (the *Bell X-1*) faster than the speed of sound. The age of supersonic flight had begun.

With every advance in transportation, the world seemed to shrink a little bit more. On October 12–13, 1992, it probably shrank to its minimum. That was when an Air France Concorde flew around the world in just 32 hours and 49 minutes.

▼

*Traveling by railroad between Britain and continental Europe was revolutionized by the opening of the Channel Tunnel in 1994. This is one of the huge machines that bored the tunnels.*

# Congestion AND POLLUTION

**Improvements in technology have made transportation faster, more comfortable, and more reliable. But so many people and goods are on the move that the road networks, shipping lanes, and airways are becoming increasingly congested.**

▲

*Aircraft wait for a take-off slot at a busy airport in Newark, New Jersey.*

designated air "highways" and becoming concentrated around airports. The busiest airports, such as Chicago's O'Hare and London's Heathrow, handle more than 60 million passengers every year. Planes take off and land every few minutes. Some shipping lanes in the sea can become packed, too. The English Channel is one of the most congested with as many as 500 ship movements a day.

Nearly 700 million vehicles speed along the roads of the world. Evenly distributed among the billions of miles of roads, these vehicles would have plenty of space. The problem is that they are mostly concentrated on a relatively small number of main roads in urban areas. The result is ever-increasing traffic congestion.

A similar situation exists for air travel, with planes flying along

## Traffic problems

Increased traffic on land, at sea, and in the air brings a number of problems. It becomes difficult to control safely in spite of modern technology, such as radar and satellite navigation.

Increased traffic also puts more strain on the environment. For example, building more roads and airports takes vast amounts of land and can disrupt the lives of people living nearby with noise and pollution.

Virtually every method of transportation on land—except the bicycle—causes pollution in some way. Air pollution creates global warming, which has become a major environmental problem. Some people believe that global warming is disrupting normal weather patterns around the globe.

## Air pollution

Most modern methods of transportation pollute the air because the engines that power them burn fossil fuels. These are fuels that come from organisms that flourished on Earth hundreds of millions of years ago. The main fuels are produced from petroleum (crude oil). Most cars burn gas, most trucks burn diesel oil, and most planes burn jet fuel.

Burning these fuels produces gases such as nitrogen, carbon oxides, and particles of unburnt hydrocarbons. In congested city streets, especially on hot days, these pollutants create a kind of smoky fog called "smog." Smog is unhealthy, particularly affecting people who suffer from respiratory illnesses such as asthma and bronchitis. Notorious in large cities across the world, smog is particularly bad in Los Angeles, Mexico City, Athens, and Bangkok.

## Oil pollution

Ships create pollution by discharging oil into the sea, either accidentally or deliberately. Most pollution happens when oil tankers are wrecked in collisions at sea or on the shore. These kinds of tanker accidents happen frequently. Late in 2002, the *Prestige* sank off the coast of Spain, the fourth major oil spill in European waters in a decade.

The released oil forms a slick that can stretch for many square miles. If an oil slick reaches shore, the effect can be devastating to the environment. It destroys the beaches and kills large numbers of seabirds and fish, as well as other marine life, such as seals.

▼

*Smog caused by air pollution is a big problem in many of the world's cities. This busy road is in Bangkok, Thailand.*

# On the ROAD

**The Romans built the world's first major road network, with a total length of more than 50,000 miles (80,000 km). It was the Romans who established the basic principles of road building.**

▲

*Many modern highways are toll roads that drivers have to pay to use. These toll booths are on the Autoroute Nord in France.*

Roman roads were built as straight as possible. They consisted of layers of stones on a foundation of hard-packed soil. The surface was curved, or cambered, to allow water to run off, and there were drainage ditches at the edges.

Modern roads are also made as straight as the terrain allows. They consist of a base of crushed stone on top of soil that has been compacted by heavy rollers. The final course, or pavement, is laid on top. This is either tarmac—a mixture of tar and crushed stones—or concrete.

A great deal of other road construction work is often needed, such as bridges and tunnels. Some of these are truly spectacular. For example, the Akashi-Kaikyo suspension bridge in Japan has a record span of 1.2 miles (1,990 m), while the

10-mile (16-km) long St. Gotthard tunnel was constructed through solid rock in the Alps.

## Superhighways

When the Roman Empire collapsed, their fine roads fell into disrepair. Not until 1,500 years later, in the 1920s, were long, straight, and well-engineered roads built again. This happened first in Italy, where these special roads—built for motor vehicles only—were called *autostrada*.

Today, most countries have their own specially built "superhighways." They are called motorways in Britain, *Autobahnen* in Germany, *autoroutes* in France, and expressways or freeways in North America. Some of these highways are toll roads that drivers have to pay to use.

In the 1920s, the United States built the world's first transcontinental road from New York to San Francisco. Known as the Lincoln Highway, it stretched for nearly 3,400 miles (5,500 km). Today, the United States boasts the biggest network of superhighways, the Interstate, which runs for nearly 45,000 miles (70,000 km).

Superhighways are designed to safely handle high-speed traffic. They have many traffic lanes, and traffic going in opposite directions is kept separated. There are no traffic lights or intersections. Minor roads pass over or under the freeways. Traffic enters and leaves at aptly named "cloverleaf junctions."

## Road traffic

Nearly 500 million cars make up the bulk of the vehicles on the world's roads today. They vary widely in size, shape, and performance. Each is designed for a specific market. There are compact cars, station wagons, hatchbacks, sporty convertibles, load-carrying pick-up trucks, minivans, and four-wheel drive SUVs (sports utility vehicles).

Most of the other vehicles on the roads—about 200 million of them—are commercial vehicles. The most common ones are trucks and buses. But there are many others, including

fire engines, garbage trucks, concrete mixer trucks, and tankers.

In most countries, trucks form a vital part of the transportation network. They carry more goods than the railroads because they are much more flexible in operation. Many trucks are articulated—split into a power unit, or tractor, and a semi-trailer that carries the goods. The two parts are connected by a swiveling coupling that makes the vehicle much easier to maneuver.

Often, trucks work with ships and railways to transport a variety of goods in standard-sized containers (*see page 28*).

▲

*An aerial view of a section of expressway in Los Angeles that shows part of a "cloverleaf junction."*

# On the ROAD

▲

*This Formula One racing car relies on spoilers to help keep it pressed down onto the surface of the track.*

## Car design

Early cars were box-like and did not travel very fast. As speeds increased, car manufacturers came up with more streamlined designs that would reduce the drag, or air resistance, upon them. Today, designers test scale models of their new cars in wind tunnels, just like aircraft designers do. They can see how streamlined the designs are by trailing streams of smoke past them.

When a car travels very fast, air gets under the body and tends to lift it off of the ground. This reduces the grip of the tires and makes for dangerous driving. Some manufacturers fit spoilers to their vehicles to prevent this. Spoilers are upside-down airfoils that develop a downward force when they travel through the air. This presses the car to the road surface.

High-speed dragsters and racing cars are fitted with spoilers on the front and back to help maintain grip on the track. Spoilers also helped *Thrust SSC* (Super Sonic Car) to grip the desert floor in Nevada when it broke the sound barrier in 1997 with a speed of 763 miles (1,228 km) per hour.

## Concept cars

Car manufacturers often introduce new ideas in car design using radical concept cars. These are usually one-of-a-kind vehicles built to demonstrate new technologies. One example is the Hy-wire that General Motors introduced at the Paris Motor Show in 2003.

The Hy-wire has a body like no other vehicle and a fuel-cell power unit. But it also has revolutionary "by-wire" control. This means that all essential functions, such as steering, accelerating, and braking, are controlled electronically, not manually.

## Safety first

On ordinary roads, cars each weighing more than a ton travel in opposite directions at speeds of 60 miles (100 km) per hour or more. If they happened to collide, they would be wrecked, and the drivers and passengers could be killed.

In fact, road deaths have become one of the world's biggest killers, with more than one million people dying in this way every year—more than 40,000 in the United States alone. Knowing that accidents will inevitably happen, car manufacturers try to design their cars to be as safe as possible.

They develop their ideas in crash tests, in which they deliberately crash cars to see the effect on dummy passengers. A big problem is how to keep the passenger compartment intact in a collision. One way is to strengthen it to form a "safety cage." To prevent the passengers from being thrown forward by the impact, cars are fitted with safety belts and airbags that inflate in a collision. Sidebars help resist sideways impacts.

## Smart innovations

A whole range of new traffic control features is being developed to make driving safer. Radar-based "intelligent" cruise control is already available. It automatically speeds up or slows down a car to maintain a safe distance from other vehicles. Waiting in the wings is a plan in which "smart" roads will be embedded with sensors, and traffic will be controlled by a computerized collision avoidance system.

Navigation satellite technology (*see page 39*) will be increasingly used to monitor traffic flow and pinpoint individual vehicles. It could be used to restrict speeds according to speed limits and keep track of which drivers need to pay road tolls.

▼

*This photograph shows a crash test in progress. An airbag has inflated to absorb the force of the impact on the dummy driver.*

# the MECHANICS

**With nearly 14,000 different parts, the modern car is a highly sophisticated machine. It is a collection of many different systems working together.**

▲

*Car bodies are welded together and finished on a car factory's highly automated assembly line.*

The body is the largest system, made from steel sheets welded together into a single shell. This design gives the body its structural strength.

The steering system guides the car. The electrical system provides electricity. The suspension system helps cushion passengers from road shocks when the car travels on bumpy roads. The braking system allows the driver to slow down and stop the car. But the most important mechanical systems are the engine and transmission. A few cars have a body built onto a separate framework, or chassis.

## Gas engines

Gas engines still power most cars. They are reciprocating engines, in which pistons move up and down in cylinders. Most work according to a four-stroke cycle, meaning that power is produced in each cylinder on one of every four strokes (movements) of the piston.

In the cycle, the fuel mixture is drawn into the cylinder, compressed, and then exploded by a spark from a spark plug. The resulting gases force the piston down to produce power. Finally, the burnt gases are forced out through the exhaust system.

A few cars have a completely different kind of gas engine called the Wankel. This is a rotary engine in which the expanding gases produced by burning fuel spin around a rotor.

## Diesels

Increasing numbers of cars (and most trucks and buses) now have diesel engines. These burn diesel fuel, which is a light type of oil. They, too, are reciprocating piston engines that work on a four-stroke cycle. But they work by what is called compression-ignition. In the diesel cycle, only air is compressed, which makes it very hot. Then diesel fuel is injected into it and explodes, forcing the piston down on its power stroke.

## Transmission

The up-and-down movement of the pistons in the cylinders of ordinary engines is converted into rotary motion by a crankshaft. At the end of the crankshaft, there is a heavy flywheel that helps smooth out its motion. The flywheel connects with the car's transmission—the units that transmit motion to the driving wheels.

The main units are the clutch, gearbox, and final drive. In cars with manual transmission, the clutch is used to disconnect the engine from the gearbox when the driver wants to change gear. The driver selects different sets of gears in the gearbox to make the car travel at different speeds for the same range of engine speeds. In cars with automatic transmission, the clutch and gearbox work automatically.

## The drive

Some cars have their driving wheels at the front (front-wheel drive), some have them at the rear (rear-wheel drive), and others use all of the wheels to drive (four-wheel drive).

In cars with rear-wheel drive, a propeller shaft carries motion from the gearbox to the final drive. This feeds motion to the wheels. Cars with front-wheel drive have the clutch, gearbox, and final drive integrated into a single unit. Using four-wheel drive provides better grip over rough terrain and in slippery conditions.

▼

*Cars are complicated pieces of machinery. In a car factory, robots assemble many of the manufactured components.*

# the mechanics –
# HYBRIDS

**Gas engines are not very efficient and are able to harness only about one-fifth of the energy in the fuel. On average, a typical medium-sized car can travel only about about 30 miles per gallon (10 km per l) of fuel. Diesel engines are much more efficient, giving nearly twice this figure.**

Cars are now being made that are capable of a staggering 100 miles per gallon (35 km per l). They are cars called hybrids and are powered by both a gas engine and an electric motor. The Honda Insight and Toyota Prius were among the first hybrids to be offered to the public.

In a hybrid car, batteries power the electric motor. They are kept charged by the engine and by energy generated in braking that would otherwise be lost.

## Reducing pollution

Gas engines pollute the air by emitting toxic fumes in their exhaust. Increasingly, governments require vehicles to emit lower amounts of pollutants.

In most new cars, low emissions are achieved by fitting a catalytic converter in the exhaust system that contains platinum and rhodium. These metals are catalysts; they promote chemical reactions that convert the harmful gases in the exhaust into harmless ones.

The gas that occurs naturally with petroleum is a much cleaner fuel than gasoline, and it is beginning to be used more in slightly modified gas engines. It is stored in the vehicle in the form of a liquid under pressure and is known as LPG (liquefied petroleum gas).

## Biofuels

Diesel engines also produce pollution, although not as much as gas engines. But a much lower emission fuel for diesels is coming into use. Called biodiesel, it is produced by refining plant oils such as olive oil, sunflower oil, and oil made from rapeseed. Biodiesel made by recycling used cooking oil is also beginning to appear on the market.

## Electric propulsion

The ultimate goal for road-vehicle emission is zero—the emission of no pollutants at all. Electric cars, popular in the early days of the automobile, can achieve this.

The majority of electric vehicles on the roads are not cars, but utility vehicles such as milk trucks. They are driven by electric motors powered by batteries. Ordinary batteries are very heavy and work for a relatively short time before they have to be recharged. Used in an ordinary car, they give poor top speed and acceleration.

The future of electric cars seems to lie in fuel cells, the power units that provide electricity in the space shuttle. In a fuel cell, hydrogen fuel combines with oxygen in the air to produce electricity. Water is formed as a waste product. There are no harmful emissions.

Fuel-cell cars are already available. In 2003, Los Angeles took delivery of the first major order of fuel-cell vehicles, the Honda FCX. And General Motors recently unveiled the amazing futuristic Hy-wire concept car that is powered by fuel cells (*see page 14*).

*Oil produced from the rapeseed plant, shown here, may be a major source of diesel fuel in the future.*

# On the RAILS

**After two centuries, railroads remain an invaluable means of transportation for both goods and people. They are capable of shifting enormous loads, make efficient use of fuel, and cause little pollution.**

▲

*Four powerful locomotives haul a train through the southwest United States.*

One reason that the railroads make efficient use of fuel is because trains have steel wheels that run along steel track. Friction between the two is low.

The twin rails that form the track are made of short sections joined together. On modern lines, the lengths are continuous, welded into lengths several miles long.

In most countries, the distance between the twin rails is 56.5 inches (1.435 m). This distance, known as "standard gauge," was the one chosen by George Stephenson in the early 1830s. But broader and narrower gauges are also used. A

▶▶

*Most diesel locomotives, such as this one, are diesel-electric units.*

narrow gauge of 40 inches (1 m) is the most common alternative gauge.

## Letting off steam

Until the middle of the last century, steam power units, or locomotives, hauled most trains. Gradually, diesel and electric locomotives began to take over because they were cleaner and much more efficient.

A few steam-powered railroads are still in regular use in Africa and Asia, but in other countries they operate only as a tourist attraction. However, there are signs that steam could one day make a comeback. Engineers are already experimenting with turbo-charged locomotives that burn pulverized (powdered) coal.

## Diesel power

Diesel locomotives haul most trains around the world. They use the same kind of diesel engine as trucks and buses. But they are much more powerful, with up to 20 times the power output of a typical heavy truck.

In a few diesel locomotives, power is transmitted to the driving wheels via a gearbox or a kind of hydraulically operated, automatic transmission that diesel road vehicles use.

But most locomotives are diesel-electric and have electric transmission. The engine is coupled to a generator to produce electricity that is then fed to electric motors that drive the wheels.

## Electric locomotion

Electricity powers the world's fastest trains, such as Japan's bullet trains and France's TGVs (*see page 25*). The power cars on electric trains receive their electricity from overhead power lines. They do so by means of a spring-loaded hinged arm (called a pantograph) on the roof.

Some power lines supply fairly low-voltage DC (direct current) electricity to the locomotive, and this is fed directly to DC motors that drive

the wheels. Most lines, however, supply AC (alternating current) electricity at high voltage, typically 25,000 volts. On board the locomotive, a transformer reduces the voltage, and a rectifier changes it to DC so it can power the electric drive motors.

## Turbine trains

Some of the most powerful locomotives have been the gas turbine-powered units used by the Union Pacific Railroad in the United States for hauling heavy freight trains through the Rocky Mountains. In turbine locomotives, the hot gases produced by burning fuel spin a turbine that drives the wheels.

Lighter, faster locomotives powered passenger trains in North America from the late 1960s to the 1980s. Some of these turbotrains have now been refurbished. The first began carrying fare-paying passengers in April 2003. On a suitable track, these trains can reach speeds up to 170 miles (270 km) per hour.

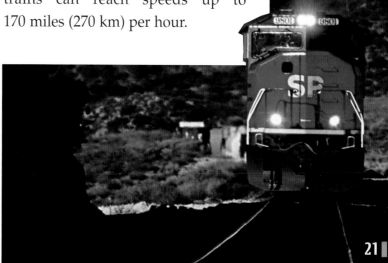

# On the RAILS

## Signaling and safety

In most countries, railroads are continually under improvement. Operating companies constantly lay new track and buy better locomotives and rolling stock. As a result, trains are getting faster and faster, and some routinely reach speeds of 150 miles (250 km) per hour or more.

For railroads to run safely, they must be carefully monitored. The days when signalmen would control individual trains along a section of track by setting signals and changeover points are long gone. Signals and points are now set from a signal center that controls traffic over hundreds of miles of track.

Commands to change signals and points travel along electronic track circuits. These also allow controllers to pinpoint the position of all of the trains.

▲

*Computers keep track of traffic at a modern railroad signaling center.*

*This straddle monorail system moves people at Darling Harbour, Sydney, Australia.*

Many trains have automatic train control that, for example, cuts off locomotive power and applies the brakes if the train runs through a red light. This eliminates driver error, the cause of many fatal crashes.

## Mass transit

Traffic congestion is increasing rapidly in most large cities, making journeys across town longer and more stressful. It was a problem in London as far back as the 1860s, when city planners decided to build a railroad under the crowded streets. The London Underground was born.

Today, many large cities have underground railroads, also called metros or subway systems. The New York Subway and the Moscow Metro are the world's busiest networks, each handling as many as six million passengers every day.

Such underground railroads pioneered what is known as mass transit—the transporting of as many passengers as possible in the shortest time. The New York and Moscow networks date back to early last century. Much more recent are the San Francisco Bay Area Rapid Transit (BART) system, the Washington Metro, and Hong Kong's Mass Transit Railway.

Most mass transit trains travel underground for only part of the time, spending the remainder on the surface or on elevated track. A section of the BART system even runs along the seabed of San Francisco Bay in an underwater tunnel.

## Monorails

Some cities possess another kind of elevated railroad. Known as a monorail, it does not run along the usual twin-rail track but along a single rail.

The world's first and most successful monorail is the Wuppertal Schwebebahn in Germany which has been operational since 1901. It is called a suspension monorail because the train hangs beneath the elevated track.

Monorails are familiar to people who have visited Disneyland amusement parks. Some monorails transport passengers and their luggage at airports. These are known as straddle monorails because they are mounted on top of the track, straddling it. Very fast monorails powered by electromagnets could also enter service soon (*see page 25*).

# SUPER railroads

**Trains that tilt to go around corners speed up journeys on ordinary railroad lines. But the fastest trains run on specially built track. Ingenious maglev trains will go even faster.**

▲

*A Japanese bullet train streaks along the specially built track of the Shinkansen with sacred Mount Fuji in the background.*

In most countries, railroad tracks follow routes originally built in the 19th century, twisting and turning to follow the landscape. However, when traveling at high-speed around bends, trains have to slow down. This is because centrifugal force tends to fling the train sideways and can throw passengers around.

To achieve higher speeds, railroad designers have come up with the tilting train. In this design, the entire body of the train leans into the bend, just like a motorcyclist does when going around a corner. A system of hydraulic rams tilts the body of the train when sensors detect that the track is curving.

One of the most successful tilting trains is the *Pendolino* ("Little Pendulum"), developed in Italy. It is in operation in Italy and in many other countries, including Britain, Canada, Germany, Malaysia, and Switzerland. Spanish Railways operate a similar design, known as the *Talgo*.

## The bullet trains

In 1964, Japan's *Shinkansen* ("New Trunk Line") ushered in a new approach to railroads. It was built from scratch as a high-speed line. Originally running from Tokyo to Osaka, the Shinkansen now extends for more than 1,250 miles (2,000 km).

The Shinkansen trains were aerodynamically designed to reduce air resistance and were tested in a wind tunnel. Their sleek shape earned them the name of "bullet trains."

The trains that run on France's high-speed rail network are also well-streamlined. They are known as TGVs (*Trains à Grande Vitesse*). They travel mostly on purpose-built track. In 1990, one reached a record speed of 320 miles (515 km) per hour, but their normal operating speed is up to about 160 miles (260 km) per hour.

## Low-flying trains

The maximum speed that a train can possibly achieve on ordinary track is about 250 miles (400 km) per hour, after which the wheels lose their grip. The high-speed trains of tomorrow will "fly" over the track, eliminating friction between train and track and thus allowing much higher speeds.

French engineers pioneered the "flying train" by using the hovercraft principle (*see page 31*). Their experimental vehicle, called the Aerotrain, glided along a monorail track on a high-pressure cushion of air. It was fast—but incredibly noisy.

## Magnetically levitated

Another kind of "flying train" has proved much more promising. It suspends itself above the track by means of a very strong magnetic field. This effect is called magnetic levitation, or maglev for short.

A maglev train carries powerful electromagnets along each side. There are more powerful magnets along the sides of the guideway (track), or underneath, depending on the design. In operation, the two sets of magnets either attract or repel each other to lift the train in the air.

Most pioneering work on maglev trains was carried out in Japan and Germany. But the first commercial maglev line went into service in Shanghai, China, in February 2003.

▲

*A train on the world's first commercial maglev railway, which links the Chinese city of Shanghai to its international airport.*

# Out at SEA

**Every day, tens of thousands of ships cross the world's oceans, carrying all kinds of cargo from country to country. They vary widely in size, design, and method of propulsion.**

▲

*Much of the world's cargo travels in standard-sized containers such as those visible on the deck of this ship.*

Ships have a greater carrying capacity than any other form of transportation, but they are very slow. Many cargo vessels travel at a speed of 13 miles (20 km) per hour, and even the fastest passenger-carrying liners manage only about 37 miles (60 km) per hour.

The biggest problem with ordinary vessels is the drag, or resistance, of the water on the hull (the main body of the ship that sits in the water). Most engine power is used to overcome this resistance rather than to propel the vessel.

## Ship design

Basic ship design still depends on a famous principle established by the Greek mathematician and inventor, Archimedes (c. 287–212 B.C.). He discovered that an object will float if it displaces (pushes aside) a weight of water equal to its own weight.

Different ships have different designs according to their use. But almost all ships are built of steel and are constructed by welding steel plates together. For extra safety, most have a double bottom—the ship's hull has a double wall with a space in between. Horizontal plates form the various decks of the ship, while vertical plates called bulkheads divide them into compartments. This design makes the hull rigid and contributes to safety because individual compartments can be sealed off if they become flooded.

Several different materials have also been used in ship construction. One is glass-reinforced plastic (GRP), which is strong, does not corrode like steel, and is non-magnetic. For this reason, it has been used to build naval minesweepers with hulls that do not explode magnetic mines. However, GRP is most widely used for building yachts and other pleasure craft. Thin, reinforced concrete called ferrocement has also been used in boat construction.

## Ship propulsion

Nearly all ships are driven through the water by propellers. The most common power unit is the diesel engine, which burns oil. It is similar to the type of engine used to power trucks and buses on land, but it is much bigger. It drives the propeller shaft either through sets of gears or by using electric motors.

Some large ships are powered by steam turbines that use high-pressure steam generated in boilers. Usually the boilers are heated by burning oil, but in some ships they are heated by a nuclear reactor. Nuclear power has not proved economical for commercial craft, but it is used in some

warships and submarines. The largest nuclear-powered ships are the U.S. *Nimitz*-class aircraft carriers, measuring 1,090 feet (333 m) long and with a deck area of nearly 4.5 acres (2 h). A few icebreakers and many naval submarines also use nuclear propulsion (*see page 29*).

In the future, ship designers may look to the past to provide more efficient propulsion for ships. They are experimenting with the use of sails as an auxiliary (extra) power source so that the fuel-burning main engines can be used less. The sails take the form of huge aerofoils that were designed in wind tunnels. Computers control their movement, positioning them at exactly the right angle to the wind. The Japanese tanker *Shin Aitoku Maru* pioneered this new technology in the 1980s.

▼
*An enormous oil-carrying tanker discharges its cargo at a refinery in North Carolina.*

# Out at SEA

### Shipshape

The outward design of ships varies greatly. In general, cargo ships have very little superstructure, or body, above the main deck level. There is usually just the bridge (from where the ship is controlled), crew accommodation, and the funnels that vent the exhaust gases from the engines.

Some general cargo vessels have lifting gear, or derricks, to handle cargo that is stowed under the deck. Container ships carry their cargo in standard-sized containers stacked on the deck. These are loaded on and off by specialized handling equipment at container ports.

The largest cargo vessels are the bulk carriers that transport cargoes such as grain, ore, and liquefied gas. The giant tankers designed to carry crude oil are the biggest of all. One example is the *Jahre Viking*, measuring 1,500 feet (458 m) long and 230 feet (69 m) across. It can carry more than half a million tons of oil.

### People carriers

The most widely-used passenger ships are ferries that operate on relatively short sea crossings such as

▼

*Many cruise ships are like floating five-star hotels. This is the* Cunard Countess.

the English Channel. These ships generally have basic accommodations with restaurants and bars, lounges, and often sleeping cabins.

Relatively fast (37 miles [60 km] per hour) Sea Cat ferries are now in service worldwide. They get their name because they are catamarans, boats with twin hulls. Built of aluminium and measuring 243 feet (74 m) long, they can carry more than 430 passengers and 80 cars. Fast hydrofoils and hovercraft ferries are also in use around the world (*see page 30*).

## Floating cities

One of the fastest-growing sectors of the travel market is people vacationing on cruise ships. The latest cruise liners are like self-contained, floating cities.

A typical cruise ship is the *Voyager of the Seas*, launched in 1999. By early 2004, it had been joined by four identical sister ships. Some 1,020 feet (310 m) long and nearly 165 feet (50 m) across, these ships each carry more than 5,000 passengers and crew. They have swimming pools and movie theaters.

Coming into operation in 2004, the *Queen Mary 2* will be the largest liner ever built. At 1,131 feet (345 m) in length, it is as long as four football fields placed end to end.

## Under the waves

Submarines are ships that travel underwater. They dive by flooding their ballast tanks with water, and they surface by forcing water out of the ballast tanks with compressed air.

Widely used by the world's navies, submarines are the deadliest of fighting ships, packing enormous firepower. The largest ones are of the Russian Typhoon class and are 560 feet (170 m) long. Armed with 20 nuclear missiles with multiple warheads, they are capable of reaching targets nearly 5,600 miles (9,000 km) away.

Many large submarines are nuclear powered. They use a nuclear reactor to heat boilers that create steam to drive the turbines to turn the propellers. Other conventional submarines use electric motors to turn the propellers underwater and a diesel engine on the surface. Electric motors also power submersibles, the small submarines used in diving and underwater research work.

▲

*This Sea Cat ferry speeds passengers between Britain and France at up to 37 miles (60 km) per hour.*

# SURFACE skimmers

**Riding on underwater wings, hydrofoil boats skim over the surface of the water, achieving speeds impossible for ordinary craft. Hovercraft travel in a different way, riding on a cushion of air.**

▲

*Hydrofoils use wing-like structures to lift themselves clear of the surface and reduce drag, thus allowing high speeds.*

Ordinary boats and ships travel slowly because of the resistance, or drag, of the water on their hulls. But marine designers have overcome this problem by building craft that operate with their hulls clear of the water.

One design is the hydrofoil craft. Hydrofoil boats are fitted with underwater "wings," or hydrofoils. When they move through the water, the foils develop lift. When the boats travel fast enough, the foils lift the hull clear of the water and reduce drag to a minimum.

## Hydrofoil designs

Hundreds of hydrofoil boats are in operation around the world, both on inland waterways and for short sea crossings. They cruise at speeds of up to 50 miles (80 km) per hour. Russia has by far the largest fleet of hydrofoil craft. Most are fitted with V-shaped, or surface-piercing foils.

One of the most advanced hydrofoil designs is the Boeing Jetfoil, which has submerged horizontal foils. It is a sophisticated craft with a fully-automated, computer-controlled system. It is powered by water-jet engines. Water is sucked in through the aft (rear) foil unit and then pumped out at high pressure through twin nozzles at the base of the hull.

## Air-cushion vehicles

Another surface skimmer—the hovercraft—is capable of traveling even faster than the hydrofoil. Hovercrafts are also called air-cushion vehicles because they skim across the surface on a cushion of air.

Hovercrafts do not look like an ordinary boat, but more like a huge rubber dingy with a flexible "skirt" around the edge. Powerful fans blow air beneath the craft, and the skirt stops the air cushion formed by the fans from leaking away too quickly.

"Pushing" propellers that face backward propel hovercrafts. They can be swiveled to steer the craft, often in conjunction with rudders on tail fins at the rear. The lifting fans and propellers are usually powered by diesel engines.

Gas-turbine engines have also been used to power the largest hovercraft, such as the SRN4 ferries that operated across the English Channel from 1968 to 2000. Nearly 190 feet (57 m) long, they could travel at speeds of up to 75 miles (120 km) per hour.

## Surface effect

Air-cushion technology has also been developed for more conventional craft, leading to the design of what are called surface-effect ships (SES). The Norwegian fast patrol boat *Skjold* is a twin-hull SES that can lift its hull on an air cushion and operate in only 3.3 feet (1 m) of water.

Even more unique are ram-wing craft. They are winged machines that fly just a few yards above the surface. These "flying boats" have wings that are angled to compress the air in front of them, creating an air cushion that helps support them. This principle is known as "ground effect." German engineers have been in the forefront of this new technology with designs such as the X-114 and the Airfish 3.

◄ ◄

*Small hovercraft boats are now available for recreational use.*

▼

*This huge SRN4 hovercraft operated on the English Channel route from 1968 to 2000.*

# In the AIR

**Supersonic airliners such as the Concorde have come and gone, and the only supersonic planes still flying are military aircraft. Today, the skies belong to subsonic airliners that carry tens of millions of passengers every year. Big "super jumbo" airliners will soon enter service as well.**

*A computer-generated image showing a prototype of the Airbus A380 double-deck airliner.*

Concorde went out of service in autumn 2003 after flying for more than 30 years. It was termed "supersonic" because it traveled at more than the speed of sound. In fact, Concorde flew even faster at about Mach 2, the term for twice the speed of sound (about 1,350 miles [2,170 km] per hour). The fastest military fighter jets, such as the Russian Mig-25 Foxbat, are even faster and can fly at three times the speed of sound (Mach 3).

In early 2004, the biggest airliner in service was the Boeing 747-400, a slightly extended version of the original Boeing 747 jumbo jet that first flew in 1969. It is a subsonic craft —one that travels at less than the speed of sound. Nearly 232 feet (70 m) long and with a wingspan of more than 211 feet (64 m), it weighs more than 300 tons (272 t) when taking off with a full load of passengers. Cruising speed is about 600 miles (970 km) per hour.

## Competition in the skies

By 2005, a super jumbo built by Europe's Airbus Industrie will become the biggest airliner in the skies. Called the A380, it is only about 7 feet (2 m) longer than the 747 but has a wingspan 50 feet (15 m) greater. It is designed with a double deck that extends the entire length of the fuselage. It could carry as many as 1,000 passengers but will probably operate with a maximum of about 600. Fully loaded, it could weigh up to about 600 tons (545 t).

Not to be outdone, Boeing engineers have come up with a design for a smaller but faster airliner. They call it the Sonic Cruiser because it will travel just below the speed of sound (up to Mach 0.98, or 735 miles [1,180 km] per hour). It will have a long fuselage, with a triangular-shaped delta wing at the rear like the Concorde. Boeing also has plans for a light and highly efficient airliner known as the 7E7, built of plastic composite materials.

## Agile fighters

There is also strong competition across the Atlantic for the next-generation of military planes. Europe is developing the Eurofighter Typhoon, designed with sophisticated computer technology to be particular-ly agile in combat. In the United States, the highly maneuverable F-22 Raptor is coming into operation. The F-35 Joint Strike Fighter (JSF) is also being developed for a variety of combat roles.

Development costs for new designs featuring the most advanced tech-nologies are very high, which is reflected in the cost of the production aircraft. The U.S. F-22 Raptor, for example, costs approximately $100 million.

## Basic design

The main parts of a plane are the fuselage (body), the wings, and the tail. The fuselage holds the passengers, crew, and cargo. The wings provide the lifting force, or lift, that supports the plane in the air. The tail acts like the flight feathers of a dart to help keep the plane flying straight and steady.

The position and shape of the wings vary from plane to plane. Low-speed planes have wings that project more or less at right angles from the fuselage. Higher-speed planes have their wings swept back at an angle. The fastest planes have sharply swept-back wings or delta wings. These designs help reduce drag as speeds increase to beyond the speed of sound.

# In the AIR

▲

*A U.S. B-2 bomber takes off. Modern aircraft rely on advanced materials such as carbon fiber and titanium for strength and lightness.*

## Swing wings

Some military planes have variable-geometry, or swing wings. For low speeds at take-off and landing, the wings stick out from the fuselage at right angles. For high speeds, the wings swing into a sharply swept-back position. The U.S. F-14 Tomcat and Russian Su-24 fighters have this ability, as does the U.S. B1b Lancer supersonic bomber.

## Testing times

Aircraft designers take years to develop a new plane. Early on, they carry out extensive testing on potential designs in a wind tunnel. They suspend scale models of their craft in the tunnel and blow air past them. The data they gather tells the designers how full-size planes should behave when they are flying.

## Jet power

Most modern planes are fitted with jet engines. In the engines, fuel is burned in compressed air to produce a stream of high-speed gases. As the gases shoot backward in a jet out of the engine nozzle, they develop an equal and opposite force (called thrust) that propels the engine—and the plane—forward.

The most widely-used kind of jet engine is the turbofan. It has a big fan in front to force air into and around the fuel-burning unit. This design makes for more efficient propulsion. Some slower transport planes and small private planes use turboprop engines, in which the jet exhaust drives a turbine to spin a propeller. Turboprops are more fuel-efficient than turbofans but do not work at high speeds. However, scientists have developed a propeller power unit called a propfan that is capable of higher speeds.

## A streamlined approach

The thrust of the engines propels a plane through the air against drag, or air resistance. So designers need to reduce drag as much as possible. Testing in wind tunnels helps select the most streamlined shape for the plane—one that the air flows around as smoothly as possible.

Engine power is more effective if

the plane weighs as little as possible. For that reason, light materials are used for construction. Most widely used are aluminium alloys such as duralumin that are light but strong. High-strength titanium may be used in the highly-stressed airframes of advanced fighters such as the Eurofighter and F-22.

Composite materials are also coming into widespread use in aircraft construction. They are plastics reinforced with glass or carbon fibers. Light and strong, these materials bind together easily. The airframe of the Lear Fan, for example, is made entirely of carbon-fiber composite.

## By stealth

Stealth planes such as the F-117A and B-2 bombers have a unique design. They are called stealth planes because they are almost invisible to radar and can creep up on targets with little chance of being detected. They have a shape like no other, with angled facets that wreak havoc with radar reflections, and they are coated with radar-absorbing material. Special exhausts reduce heat emission from the engines.

◀

*This close-up view of the front of an F-117 Stealth fighter shows its radar-beating angular design.*

# Vertical TAKE-OFF

The helicopter is the most versatile of all flying machines. It can not only take off and land straight up and down, but also hover like a hawk and fly sideways, backward, as well as forward.

With few exceptions, ordinary fixed-wing planes need to take off and land on a runway up to two miles (3 km) long. However, some aircraft can operate without a runway and can take off and land vertically. They are called VTOL (vertical take-off and landing) craft.

The only successful VTOL fixed-wing plane has been the Harrier "jump jet." It achieves vertical operation by what is termed vectored thrust. This means that the plane's pilot can change the direction of the jet exhaust from its engine by means of swiveling nozzles. Swiveling the nozzles to point downward lifts the plane vertically for take-off. In the air, the nozzles swivel to send the jet exhaust backward, and the plane flies normally.

## Rotary wings

However, the most common kind of VTOL craft is the helicopter. Helicopters are widely used by the military for carrying assault troops and as gunships. Rescue services use them for air-sea and mountain rescue, and their use is widespread in business and tourism as well.

Helicopters are known as rotary-wing craft (as opposed to fixed-wing planes). This is because the rotating blades, or rotor, mounted on the helicopter act like a wing and support the craft in the air. The blades have a similar aerofoil shape to a plane's wing. When they travel through the air, they provide a lifting force, or lift.

The rotor is powered by a gas-turbine engine. It not only provides lift, but also propulsion. The rotor blades are angled so that they act in much the same way as a propeller.

## Action, reaction

In line with Isaac Newton's (1642–1727) third law of motion, the rotor blades rotate in one direction, and the body of the helicopter tends to rotate in the opposite direction. So most helicopters have a second small rotor mounted sideways on the tail. When it spins around, it produces counter-thrust that stops the body from rotating.

Some helicopters do not need tail rotors. They have twin main rotors rotating in opposite directions, so the turning effects on the helicopter body cancel out.

## A new age of airships?

The original VTOL craft was the hot-air or gas-filled balloon. Early last century, powered balloons, or airships, pioneered long-distance air travel.

Today, interest in airships is increasing now that ordinary aircraft are becoming more and more expensive to build and operate. A few airships (such as Goodyear) are already flying—they provide a stable platform for items such as aerial TV cameras. But airships could have many other uses, such as transporting cargo and construction equipment over rough, inhospitable terrain. Passenger airships could also become popular with tourists.

Modern airships have come a long way in their design. In particular, they are much safer being filled with helium instead of hydrogen. They also benefit from the latest technology in structural materials and engine design.

◄ ◄

*A Russian military helicopter discharges its cargo in a remote area.*

▼

*Modern airships such as these are filled with the non-flammable gas helium.*

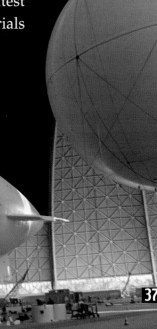

# Navigation

Radar remains an invaluable aid to navigation for ships at sea, on aircraft, and at airports. Sonar is also useful for ships and submarines. But navigation satellites have ushered in a revolution, enabling navigators to pinpoint their location anywhere on Earth to within a few feet.

▲

*The inside of the sonar room on a U.S. Navy submarine. Sonar uses sound waves to pinpoint the precise location of underwater objects.*

Navigating means finding your way across land, at sea, or in the air. Early navigators used the position of the sun by day and the stars by night to guide them. But this celestial navigation would not work in cloudy conditions. However, radio waves can penetrate clouds, and today, navigators use them to find their way around.

## Bouncing back

The most widely-used radio aid to navigation is radar. It was developed in the 1940s to detect enemy aircraft. It works by bouncing radio waves off of objects. The word stands for "**ra**dio **d**etection **a**nd **r**anging." Radar uses very short radio waves called microwaves. They are transmitted in short pulses from a rotating aerial that picks up a reflected signal, or echo, if it strikes an object.

From the time taken for the echo to be received back, the distance to the object is determined. The locations of objects are displayed on the radar screen as luminous spots.

Radar still plays a major role in preventing collisions in busy shipping lanes, particularly in bad visibility. It is also vital for air-traffic control around airports, where dozens of planes are in transit—some coming in to land, some taking off, and others moving along the runways and taxiways.

## In the water

Sonar is a similar system that uses sound waves instead of radio waves. Ultrasonic waves are used that are too high-pitched for the human ear to detect. Ships and submarines transmit the ultrasonic waves into the water and listen for echoes—sound travels fast in water. In nature, bats and dolphins use sonar for "echo-location" to navigate and to find food.

Military submarines that may stay submerged for months at a time use a system known as inertial navigation to guide them through the oceans and still remain hidden. It uses gyro-scopes to sense every change in direction and accelerometers to sense every change of speed. From this data, a computer determines the submarine's position and course.

## Satellite technology

Purpose-built navigation satellites ("navsats") are the latest aids to navigation. There are two networks of navsats circling the globe. The United States provides one called the Global Positioning System (GPS).

Russia provides the other, called the Global Orbiting Navigation Satellite System (Glonass). The European Space Agency is developing another network, called Galileo, that is due to become operational after 2008. All three systems are designed to operate in a similar way.

The GPS has 24 satellites with four each circling in different orbits about 12,400 miles (20,000 km) above Earth. (Glonass also has 24, while Galileo will have around 30.) The satellites carry very accurate atomic clocks and continuously broadcast their exact location and the precise time. A GPS receiver on a ship can then pick up signals from at least four satellites to determine its own exact location.

*Satellite navigation systems are now available for use in domestic cars.*

# Journey into SPACE

**Space shuttles will continue to carry astronauts into orbit for several years. Then new orbital space planes will begin to take over. In the long run, new technologies will be needed for journeys to other planets.**

Since 1981, U.S. astronauts have journeyed into space aboard the space shuttle. Currently, they use the shuttle to visit the International Space Station (ISS). The shuttle is a reusable launch system, in which most parts can be used over again.

The shuttle crew flies into space inside a winged orbiter. On the launch pad, the orbiter sits on a huge tank that carries fuel for its engines. Strapped to the sides of the tank is a pair of solid-fuelled rocket boosters. The shuttle is thrust into the sky by all of the engines and boosters firing together. Then, the boosters and tank separate in turn, leaving the orbiter to reach orbit. After the shuttle's mission is finished, the orbiter returns to Earth as a glider and lands on an ordinary runway.

## Future space planes

Over the years, many other designs for space planes have been put forward. Some have concentrated on what are called single-stage-to-orbit (SSTO) craft. They are self-contained, airliner-like craft that use advanced air-breathing engines in the atmosphere and rockets in space.

NASA's early concept SSTO plane, the X-30, was nicknamed the "Orient Express" because it would have been able to fly from Washington to Tokyo in less than two hours.

But the X-30 project was soon cancelled, as were other advanced concept craft, such as the X-33 and X-38. The X-38 was to be a crew-return vehicle (CRV) for the ISS—to carry the crew safely to Earth during an emergency.

## Orbital space planes

By 2003, the pressing need for a CRV and a desire for a simpler crew launch system led to NASA funding the development of an OSP, or orbital space plane. Initial work concentrated on using Boeing's X-37 craft as an advanced technology demonstra-

tor. The OSP will not have engines like the shuttle orbiter but will be launched atop an existing heavy-launch vehicle such as the Delta 4. It will have a lifting-body design so that it can be maneuvered after re-entry for a runway landing.

This system is similar to the European space plane project called Hermes that was cancelled, and one called HOPE that is being developed by the Japanese.

## Interplanetary flight

After the ISS is fully operational, space scientists will begin to look further afield in their quest to push back the space frontier. First, they will return to the Moon to set up a permanent base. This could be achieved with existing technologies, using the ISS as a construction site and spaceport.

The next push into space would be much more challenging—a journey to the planet Mars. For such a journey, which would last at least two years, a new kind of propulsion would be needed, possibly nuclear. A small nuclear reactor would provide the energy to propel charged ions to create a rocket exhaust. Ion engines are already in use and seem set to become the power source of the future.

◄ ◄

*This computer-generated image shows NASA's latest concept for an orbital space plane (OSP), the X-43C.*

# GLOSSARY

### air-cushion vehicle
A vehicle that moves over a surface on a layer of high-pressure air.

### autobahn
The German name for a high-speed road.

### autoroute
The French name for a high-speed road.

### autostrada
The Italian name for a high-speed road.

### bullet train
A highly-streamlined, fast train, such as the Japanese *Shinkansen*.

### drag
The resistance vehicles and vessels experience when they travel through air or water.

### emissions
The fumes emitted by engines.

### expressway
The North American name for a high-speed road.

### freeway
A high-speed road that is free to use.

### fuel cell
A kind of battery that produces electricity by using a fuel such as hydrogen.

### GPS
Global Positioning System; a navigation system using a network of satellites.

### hovercraft
A kind of air-cushion vehicle.

### hydrofoil boat
A boat that lifts out of the water on underwater "wings."

### jet engine
An engine that produces power by expelling a stream of hot gases.

### jumbo jet
A huge airliner that can carry a large number of passengers.

### locomotive
The power unit that hauls a train.

### maglev
Magnetic levitation; maglev trains use magnetic effects to support them.

### mass transit
A method of transporting large numbers of passengers.

### metro
A name for an underground railroad.

### monorail
A railroad track that consists of a single rail.

### navigation
Finding the way on land, at sea, or in the air.

## nuclear energy
Energy that comes from the nucleus (center) of atoms. It is used as a heat source to power some ships and submarines.

## pollution
The poisoning of the environment, such as the air and sea.

## propeller
A spinning, bladed device used to propel ships and some aircraft.

## radar
Radio detection and ranging; a method of navigation that works by bouncing radio waves off of objects.

## rotor
Something that rotates, especially the rotating blades of a helicopter.

## satellite
A spacecraft that circles around Earth.

## smog
A kind of smoky fog that forms over polluted cities.

## sonar
A method of underwater navigation that works by bouncing ultrasonic waves off of objects.

## submarine
A vessel that travels underwater.

## submersible
A small submarine used in diving operations and underwater research.

## subsonic
Traveling at less than the speed of sound.

## subway
A name for an underground railroad.

## supersonic
Traveling at more than the speed of sound.

## TGV
*Train à Grande Vitesse*; a fast French train.

## toll road
A road that drivers have to pay a fee to use.

## transmission
The part of a vehicle that transmits motion from the engine to the driving wheels.

## turbofan
The main kind of jet engine used in airplanes.

## turboprop
A jet engine that drives a propeller.

## ultrasonic waves
Sound waves too high-pitched for humans to hear.

## VTOL
Vertical take-off and landing; also the name for a craft such as a helicopter.

## wind tunnel
Apparatus in which designers study the flow of air around scale models of planes or cars.

# 21ˢᵗ-century SCIENCE

# INDEX

## A

aerofoil 14, 27, 37
air-cushion vehicle 30–31, 42
aircraft 10, 32–37, 38, 43
  design 14, 33, 34
  military 9, 32, 33, 34–35, 36
  stealth 35
airliner 9, 32–33
  subsonic 32
  super jumbo 32, 33, 42
  supersonic 9, 32
airplane 8, 10, 43
  rocket 9
airship 37

## B

biodiesel 18, 19

## C

car 8, 9, 13, 14–15, 39
  design 14–15
  electric 19
  engine *see* engine
  fuel-cell 19
  hybrid 18–19
  mechanics 16–19
  safety 14–15
Concorde 9, 32, 33
congestion 10, 22

## D

drag 14, 25, 26, 30, 33, 34, 42

## E

engine 9, 11, 16, 18
  diesel 17, 18, 21, 27, 29, 31

gas 9, 16, 18, 37
gas turbine 31, 37
jet 9, 34, 36, 42
rotary 17
steam 8
turbofan 34, 43
electric motor 18, 19, 21, 27, 29
electromagnet 23, 25
emission 18, 19, 35, 42

## F

ferry 28–29, 31
fuel 11, 16, 18, 20, 34, 40
  diesel 17, 19, 20, 21
  fossil 11
  gas 11, 18
fuel cell 14, 19, 42

## H

helicopter 36, 37, 43
hovercraft 29, 30, 31, 42
hydrofoil 29, 30, 31, 42
Hy-wire 14, 19

## I

ISS (International Space Station) 40, 41

## L

LPG (liquefied petroleum gas) 18
locomotive 8, 20, 22, 42
  diesel 20, 21
  diesel-electric 20, 21
  electric 20, 21

gas-turbine 21
steam 20

## M

monorail 22, 23, 25, 42

## N

NASA 41
navigation 8, 38–39, 43
  satellite 10, 15, 38, 39, 42
navsat 39, 42
nuclear power 27, 29, 41, 43

## O

oil 11, 17, 27, 28
OSP (orbital space plane) 41

## P

petroleum 11, 18
pollution 10–11, 18, 19, 20, 43
propeller 9, 17, 27, 29, 31, 37, 43

## R

radar 10, 15, 35, 38, 43
railroad 8, 9, 13, 20–25
  safety 22
road 8–16, 19

## S

satellite 8, 10, 43 *see also* navigation satellite
ship 8, 10, 11, 26–31,

38, 39, 43
  container 28
  cruise 28, 29
  design 26, 27, 28, 30
  surface effect (SES) 31
smog 11, 43
sonar 38, 39, 43
space 40–41
  plane 40, 41
  shuttle 19, 40
spoiler 14
submarine 27, 29, 38, 39, 43

## T

track 8, 20–25, 42
train 8, 20, 22–25
  Aerotrain 25
  bullet 8, 21, 24, 42
  maglev 24, 25, 42
  TGV 21, 25, 43
  tilting 24
transmission 16, 17, 21, 43

## U

underground railroad 22, 23, 42

## V

VTOL (vertical take off and landing craft) 36, 37, 43

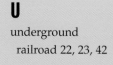